OUR AMAZING EARTH

Patricia Armentrout

A Crabtree Seedlings Book

Table of Contents

Earth's Layers 4
Earth's Atmosphere 12
Glossary 23
Index 23

Earth's Layers

We live on Earth's **crust**.

The crust is our planet's outer layer.

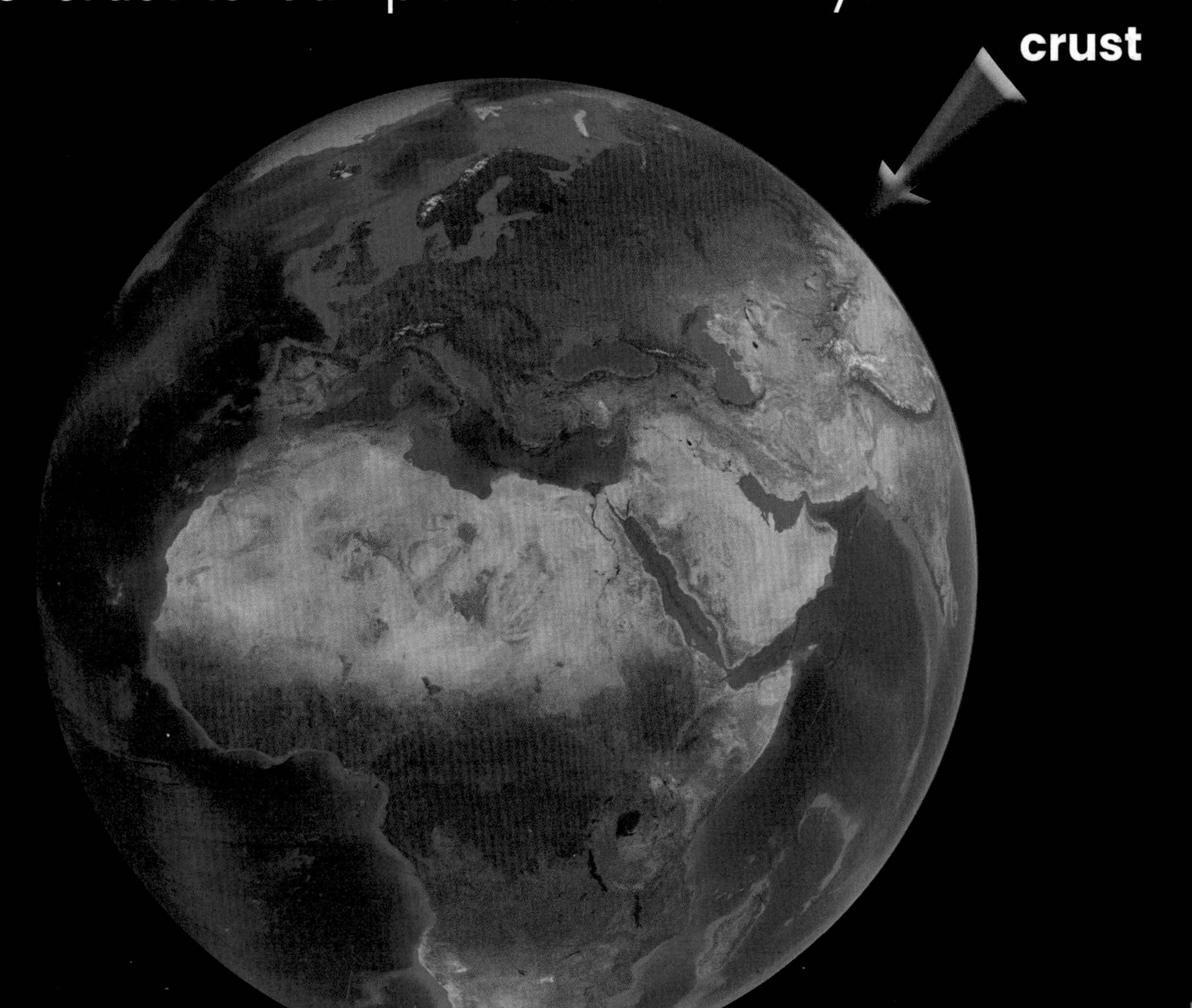

Earth has many layers.

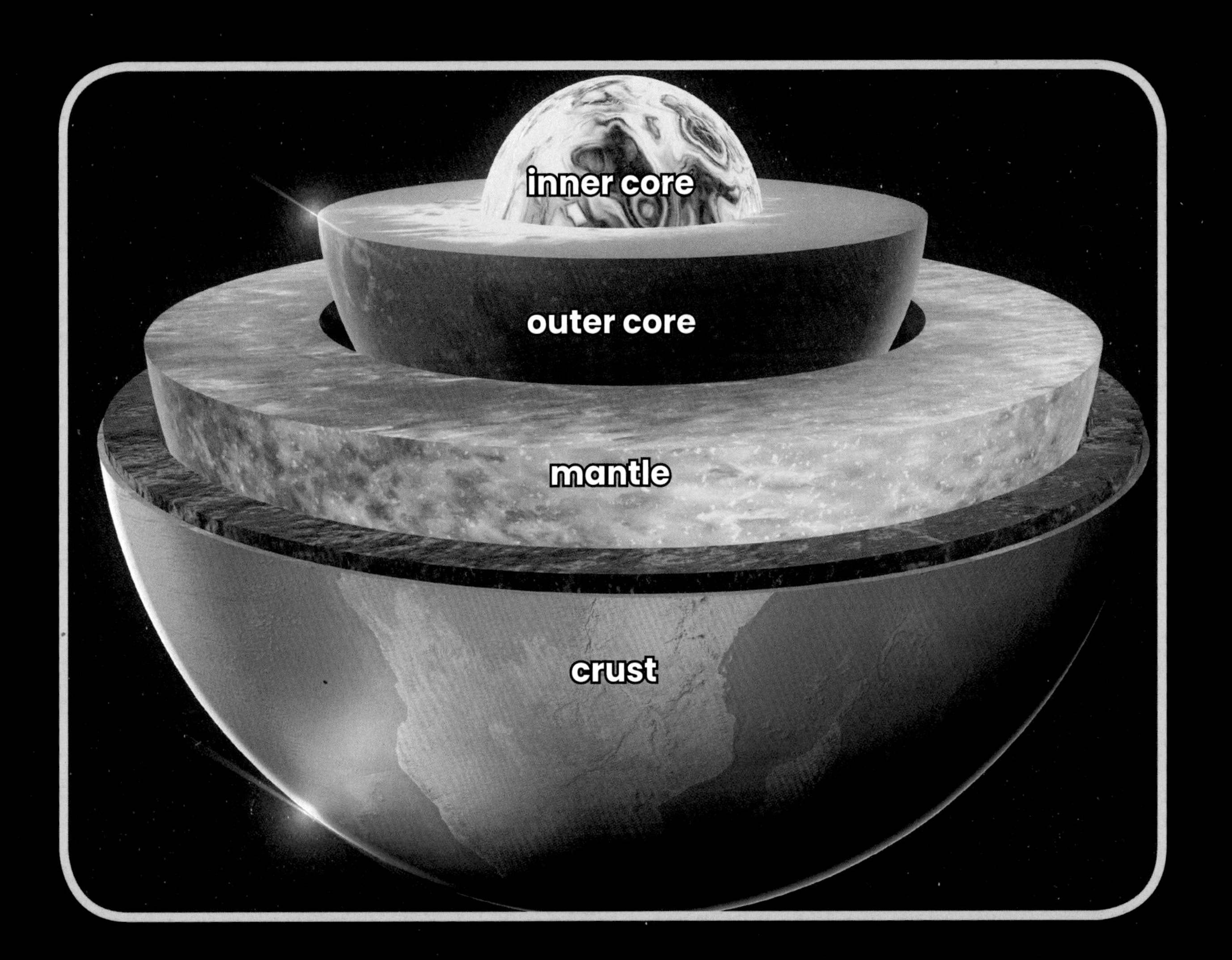

Earth's crust is made of soil, sand, and rock.

We find soil in fields.

We find sand at the beach.

We find rocks in the mountains.

field

beach

mountains

Miles below the crust is the **mantle**. The mantle is Earth's biggest layer.

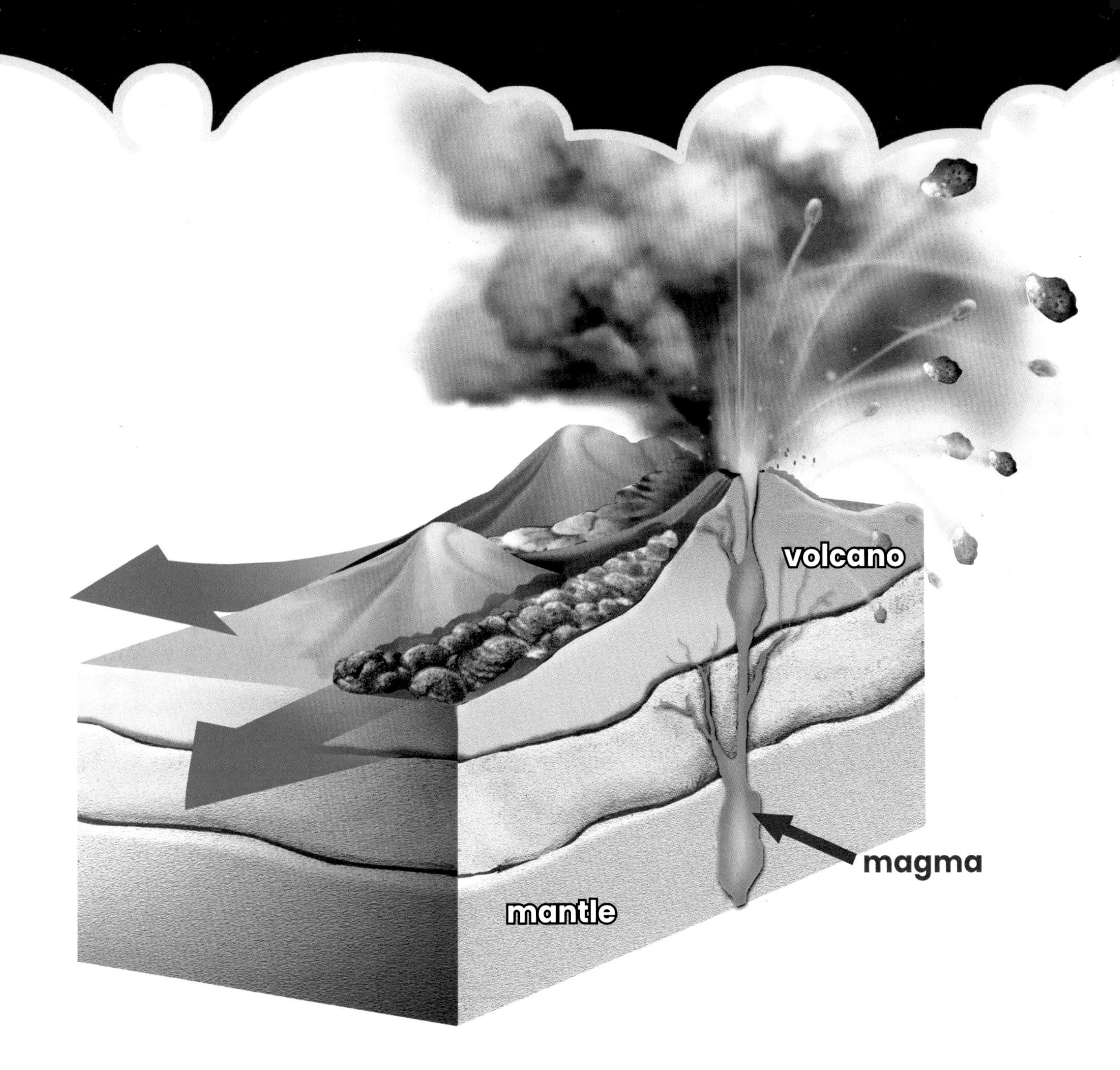

Hot **magma** flows up from the mantle, through **volcanoes** on the crust.

Deep inside the mantle are Earth's outer core and inner core.

The outer core is **liquid**.
The inner core is **solid**.

Earth's Atmosphere

Above Earth's crust is the **atmosphere**. The atmosphere has many layers.

The layer we live in has air for us to breathe. We need air.

Without air, there would be no people, plants, or animals.

We also need the Sun. Sunlight warms the air, crust, and water.

This causes weather changes like wind and rain.

Wind and rain break down rock and soil, causing **erosion**.

Over time, wind and rain help shape the land on Earth.

Above Earth's atmosphere is outer space.

atmosphere

Space is where Earth revolves around the Sun.

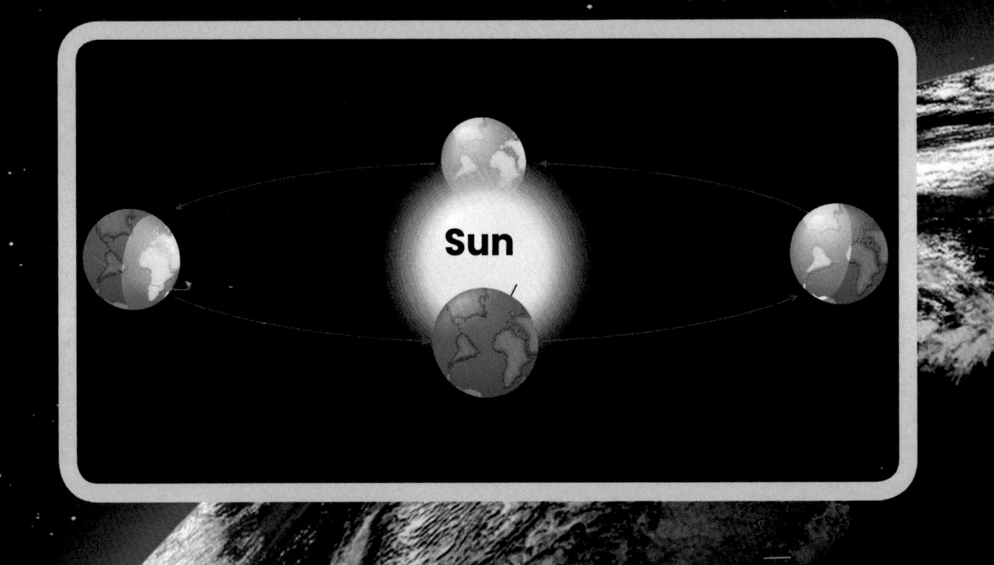

GLOSSARY

atmosphere (AT-muhs-fihr): The mix of gases around a planet

crust (KRUHST): The hard outside layer of Earth where people, plants, and animals live

erosion (i-ROH-zhuhn): Wearing away of land and rock by wind and water

liquid (LIK-wid): Wet substance that flows and you can pour

magma (MAG-muh): Very hot melted rock found deep inside Earth

mantle (MAN-tuhl): Layer of Earth between the crust and the core

solid (SOL-id): Hard, firm material that is not a liquid or a gas

volcanoes (vol-KAY-nohz): Openings in Earth's crust where magma, gases, rock, and ash can escape

INDEX

atmosphere 12, 20, 21
core 5, 10, 11
crust 4, 5, 6, 8, 9, 12, 16
erosion 18
mantle 5, 8, 9, 10
Sun 16, 22
weather 17
volcanoes 9

School-to-Home Support for Caregivers and Teachers

This book helps children grow by letting them practice reading. Here are a few guiding questions to help the reader build his or her comprehension skills. Possible answers appear here in red.

Before Reading

- **What do I think this book is about?** *I think this book is about many cool facts about Earth. I think this book is about the center of Earth.*
- **What do I want to learn about this topic?** *I want to learn how volcanoes form. I want to learn how old Earth is.*

During Reading

- **I wonder why...** *I wonder why Earth has so many different layers. I wonder why there are volcanoes.*
- **What have I learned so far?** *I have learned that hot magma flows up from the mantle, through volcanoes on the crust. I have learned that above Earth's crust are the many layers of the atmosphere.*

After Reading

- **What details did I learn about this topic?** *I have learned that without air there would be no people, plants, or animals. I have learned that Earth's crust is made of soil, sand, and rock.*
- **Read the book again and look for the glossary words.** *I see the word **atmosphere** on page 12, and the word **erosion** on page 18. The other glossary words are found on page 23.*

Library and Archives Canada Cataloguing in Publication

Title: Our amazing Earth / Patricia Armentrout
Names: Armentrout, Patricia, 1960- author.
Description: Series statement: Science in my world: level 1 | "A Crabtree seedlings book". | Includes index.
Identifiers: Canadiana (print) 20210204389 | Canadiana (ebook) 20210204397 | ISBN 9781427160515 (hardcover) | ISBN 9781039600058 (softcover) | ISBN 9781039600126 (HTML) | ISBN 9781039600195 (EPUB) | ISBN 9781039600263 (read-along ebook)
Subjects: LCSH: Earth (Planet)—Juvenile literature. | LCSH: Earth sciences—Juvenile literature.
Classification: LCC QE29 .A76 2022 | DDC j550—dc23

Library of Congress Cataloging-in-Publication Data

Names: Armentrout, Patricia, 1960- author.
Title: Our amazing earth / Patricia Armentrout.
Description: New York : Crabtree Publishing, [2022] | Series: Science in my world : level 1 - a Crabtree seedlings book | Includes index.
Identifiers: LCCN 2021018997 (print) | LCCN 2021018998 (ebook) | ISBN 9781427160515 (hardcover) | ISBN 9781039600058 (paperback) | ISBN 9781039600126 (ebook) | ISBN 9781039600195 (epub) | ISBN 9781039600263
Subjects: LCSH: Earth (Planet)--Juvenile literature.
Classification: LCC QB631.4 .A76 2022 (print) | LCC QB631.4 (ebook) | DDC 550--dc23
LC record available at https://lccn.loc.gov/2021018997
LC ebook record available at https://lccn.loc.gov/2021018998

Crabtree Publishing Company
www.crabtreebooks.com 1–800–387–7650

Print book version produced jointly with Blue Door Education in 2022

Written by Patricia Armentrout
Print coordinator: Katherine Berti
Printed in Canada/112022/CPC20221107

Photo Credits: www.shutterstock.com, www.istock.com. Cover: Earth © LOURDU PRAKASH XAVIER volcano © Steinhagen Artur. Title page and page 20-21 © muratart, Page 2-3 © Valentin Valkov. page 4 Earth © Tyler Boyes, page 5, 8 and 10-11 Earth's layers © Naeblys, page 6 soil © carroteater, rocks © Dino Osmic, mountain © Loreta Magylyte, field © AZP Worldwide, beach © DPiX Center; page 12-13 photo © Ratsamee, page page 13 and 22 illustrations © Designua; page 14 © Christopher Futcher, page 15 © Hung Chung Chih; page 16-17 © Monika23; page 18-19 © Jim Feliciano, Elena_Suvorova.

Published in the United States
Crabtree Publishing
347 Fifth Ave.
Suite 1402-145
New York, NY 10016

Published in Canada
Crabtree Publishing
616 Welland Ave.
St. Catharines, Ontario
L2M 5V6